☐ Procedure for computing the rank of a matrix, p. 287

☐ Procedure for computing the transition matrix from one basis to another basis, p. 298

☐ Gram–Schmidt process, p. 310

☐ Procedure for finding the QR-factorization of an $m \times n$ matrix, p. 328

☐ Procedure for computing the least squares solution to $A\mathbf{x} = \mathbf{b}$, p. 331

☐ Procedure for computing the least squares line for n given data points, p. 335

☐ Procedure for computing the least squares polynomial for n given data points, p. 338

☐ Procedure for diagonalizing a matrix, p. 361

☐ Procedure for diagonalizing a symmetric matrix by an orthogonal matrix, p. 369

☐ Procedure for obtaining the general solution to $\mathbf{x}' = A\mathbf{x}$, where A is diagonalizable, p. 388

☐ Procedure for identifying a nondegenerate conic section whose graph is not in standard form, p. 418

☐ Procedure for computing the matrix of a linear transformation $L: R^n \rightarrow R^m$, p. 456

☐ Random iteration approach, p. 477

☐ Procedure for solving a linear programming problem geometrically, p. 497

☐ The simplex method, p. 515

INTRODUCTORY
LINEAR ALGEBRA
WITH APPLICATIONS